The Number Two

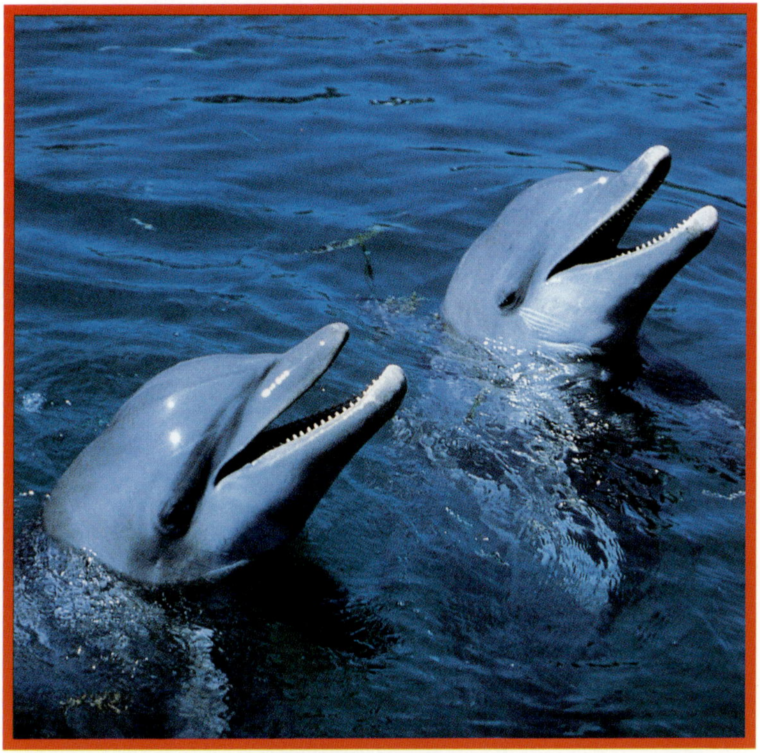

Adria Klein

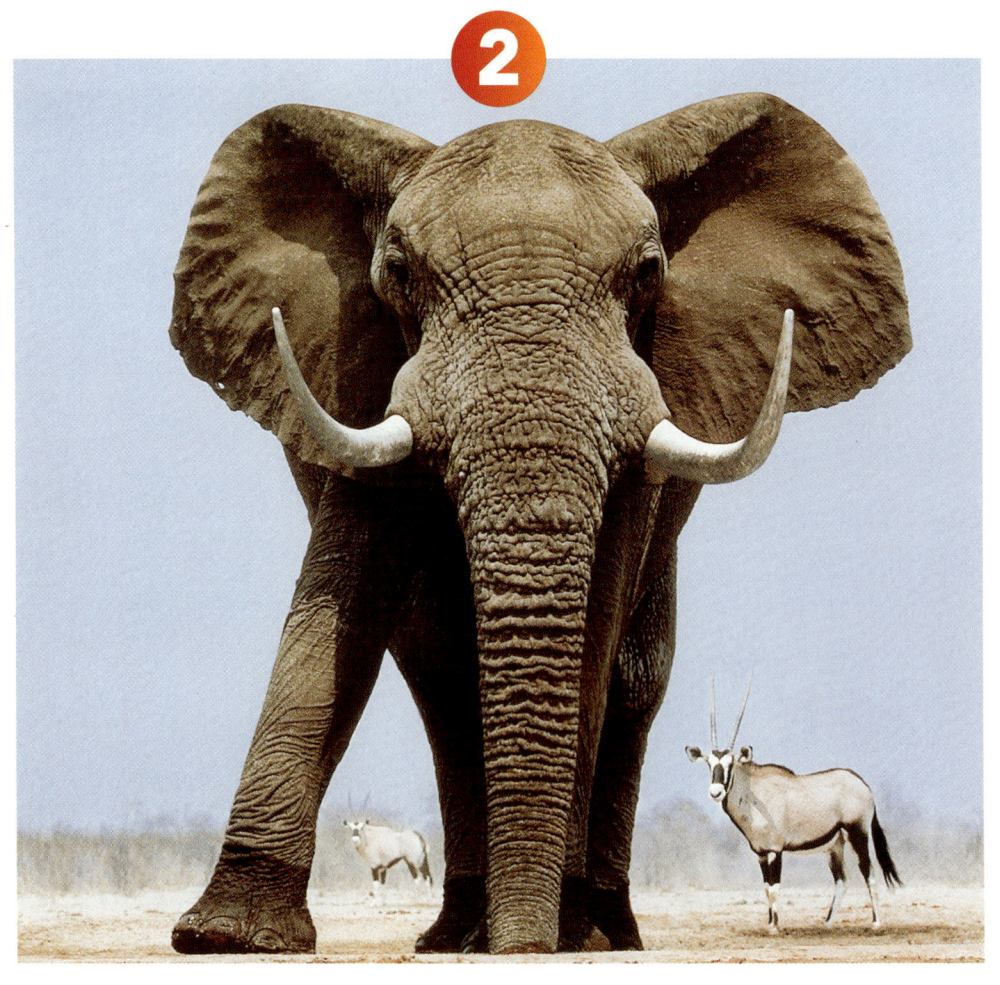

Look at my two gray ears.

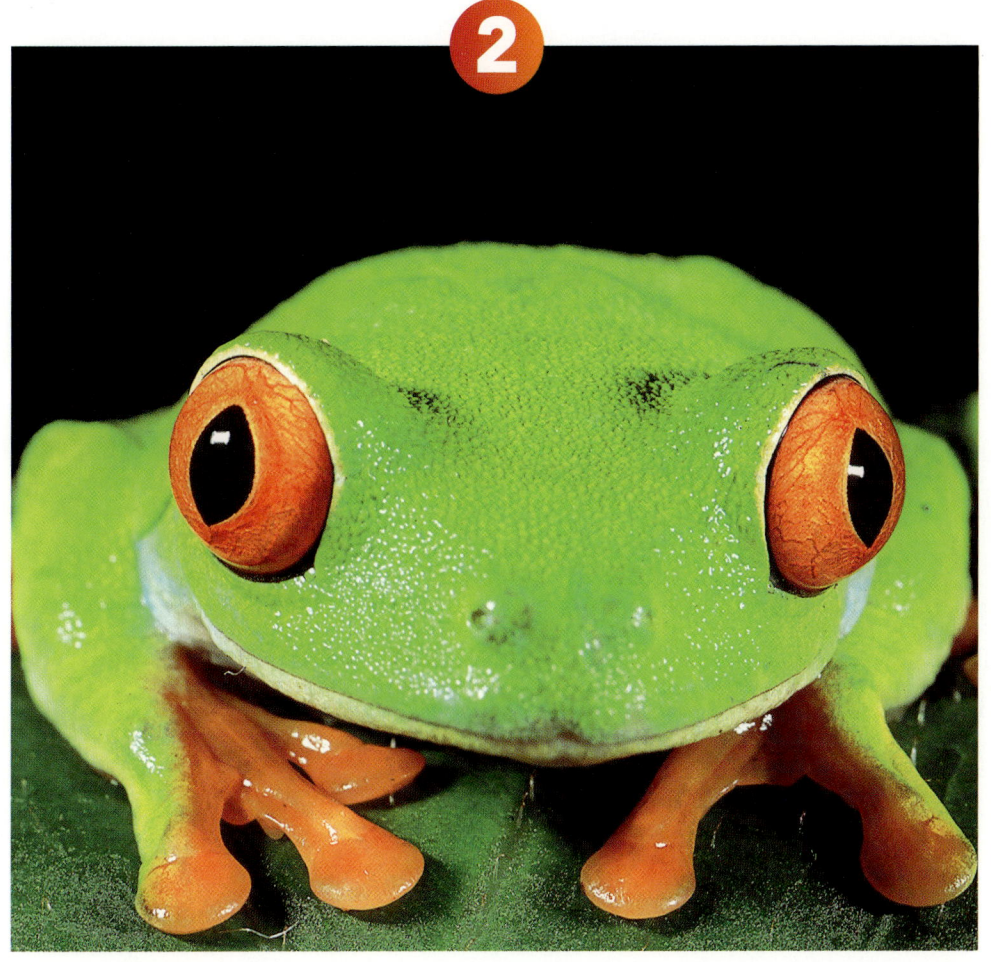

Look at my two orange eyes.

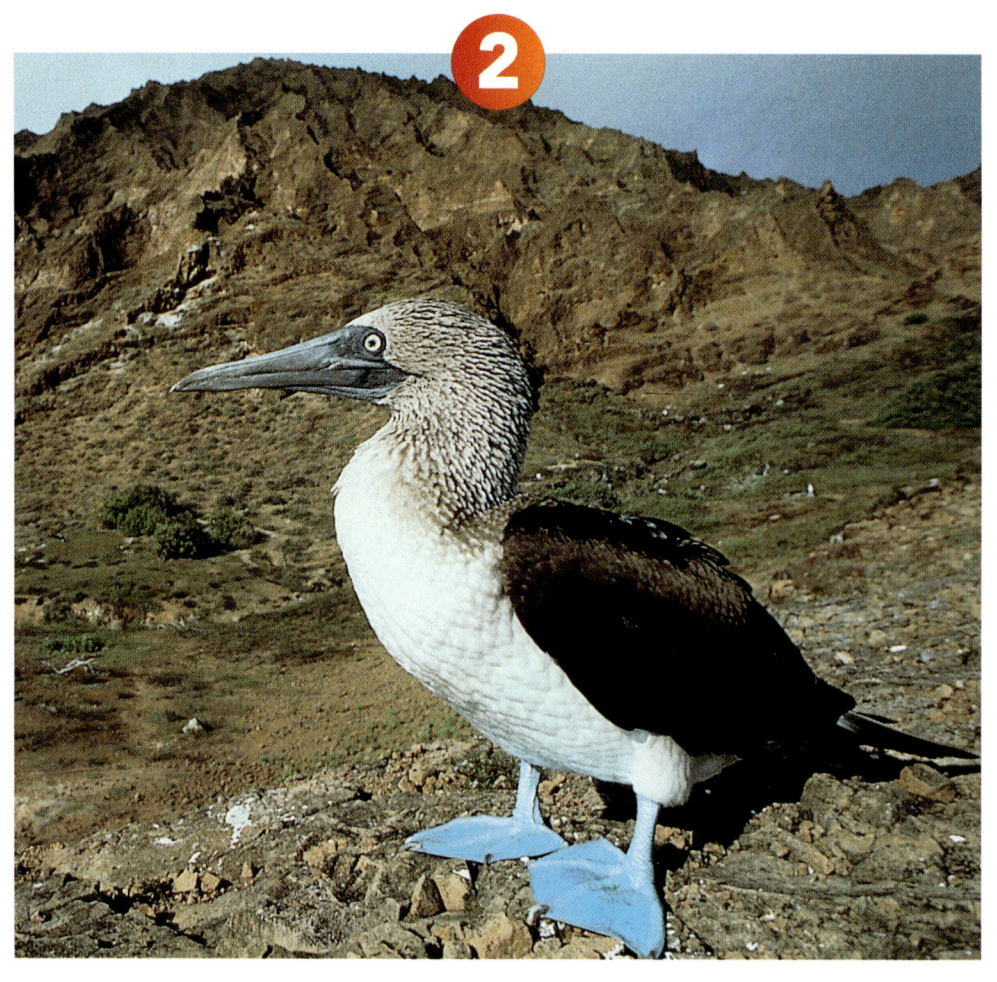

Look at my two blue feet.

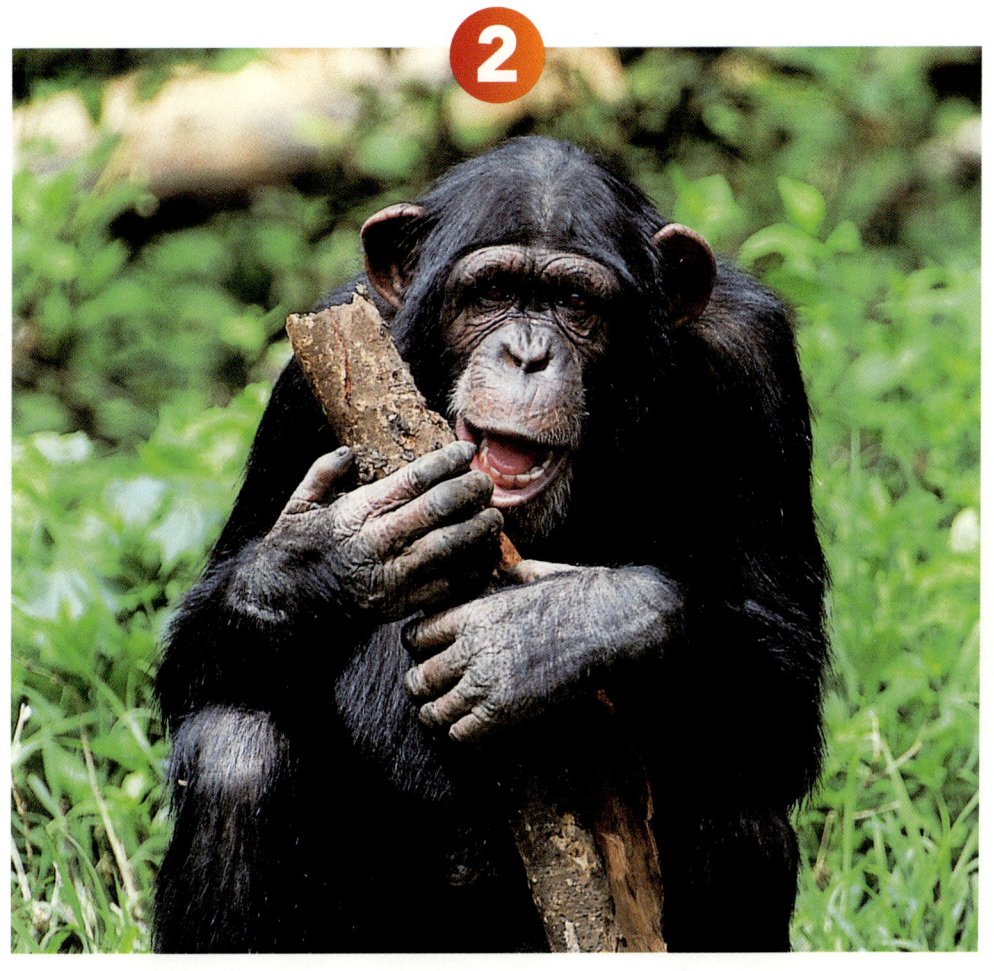

Look at my two big hands.

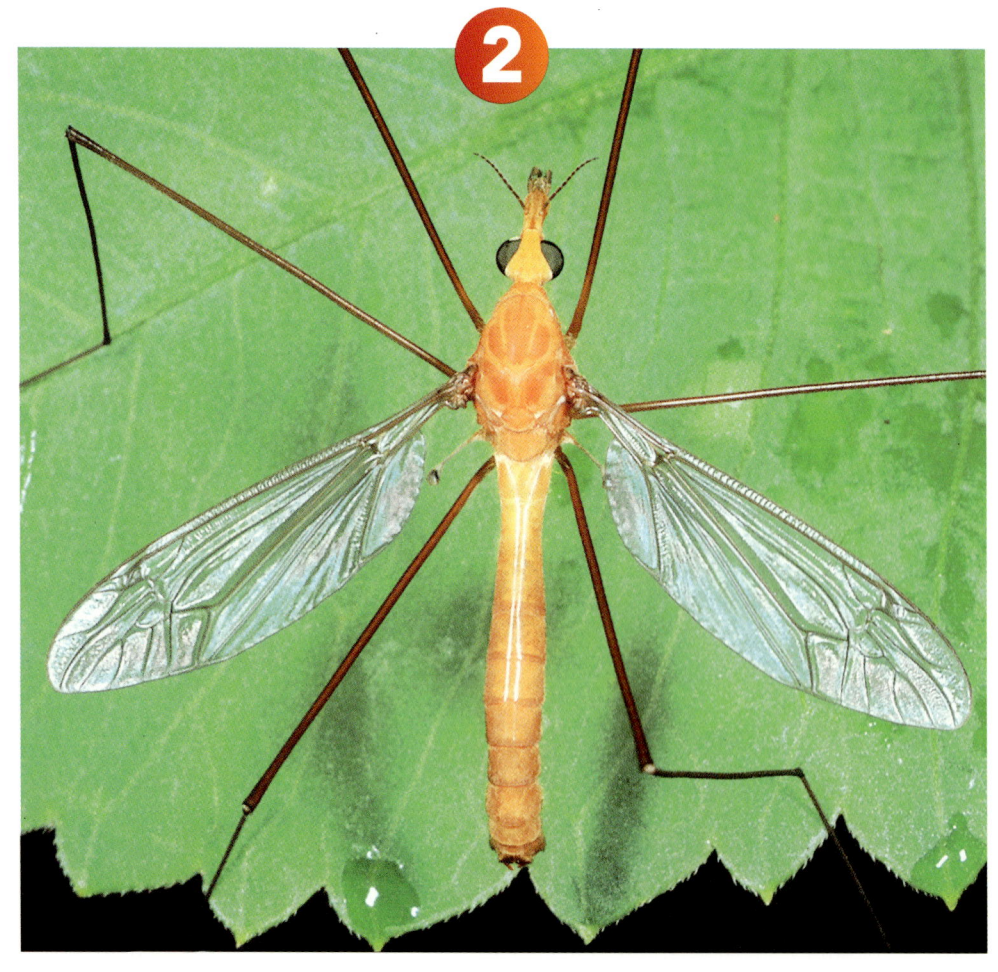

Look at my two long wings.

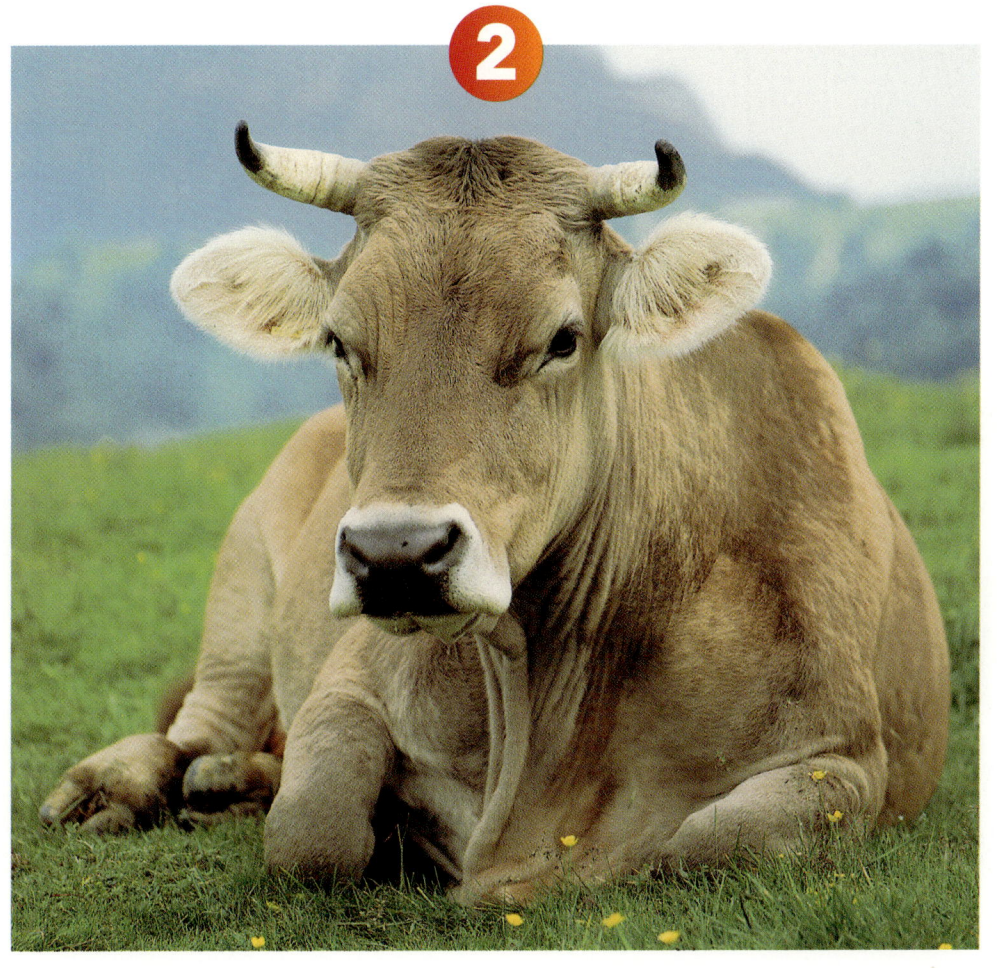

Look at my two sharp horns.

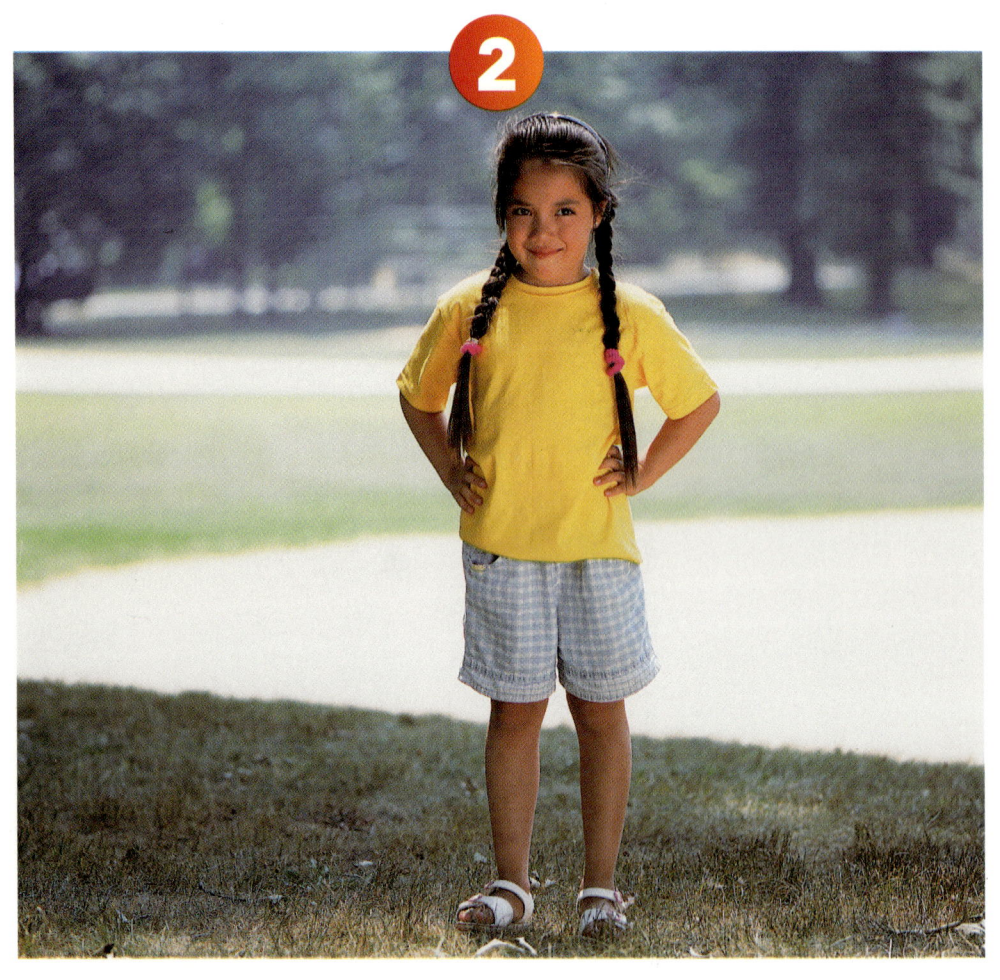

What do you have two of?